Erick Ferreira de Sousa
Wesley de Oliveira Santos
Joel Medeiros Bezerra

Use of Underground Dams in the Municipality of Bom Sucesso/PB

Erick Ferreira de Sousa
Wesley de Oliveira Santos
Joel Medeiros Bezerra

Use of Underground Dams in the Municipality of Bom Sucesso/PB

Principles and techniques

ScienciaScripts

Imprint
Any brand names and product names mentioned in this book are subject to trademark, brand or patent protection and are trademarks or registered trademarks of their respective holders. The use of brand names, product names, common names, trade names, product descriptions etc. even without a particular marking in this work is in no way to be construed to mean that such names may be regarded as unrestricted in respect of trademark and brand protection legislation and could thus be used by anyone.

Cover image: www.ingimage.com

This book is a translation from the original published under ISBN 978-3-330-76245-9.

Publisher:
Sciencia Scripts
is a trademark of
Dodo Books Indian Ocean Ltd. and OmniScriptum S.R.L publishing group

120 High Road, East Finchley, London, N2 9ED, United Kingdom
Str. Armeneasca 28/1, office 1, Chisinau MD-2012, Republic of Moldova, Europe
Managing Directors: Ieva Konstantinova, Victoria Ursu
info@omniscriptum.com

Printed at: see last page
ISBN: 978-620-8-38735-8

"Then I will give the rain of your land in its season, the early ***rain*** and the latter ***rain***, so that you may gather your grain and your must and your oil.

And I will give grass in your field to your *animals, and you shall eat and be satisfied. "*

Deuteronomy 11: 14-15

SUMMARY

Living with the effects of drought is a challenge, especially for residents of arid and semi-arid regions where this phenomenon is present and routine. The Brazilian northeast is well acquainted with this reality. In an attempt to alleviate the population's situation, actions to combat drought have emerged, but it is important that, in addition to combating drought, there is also a vision of living with it, since drought is a certainty for at least a few months of the year. The underground dam has emerged as an alternative for people to live with drought. This work was carried out in two stages: the first studied factors related to the construction of underground dams, such as: construction and location techniques, their main models, their economic viability, the results in various locations where they have been used, proper management to avoid the salinisation process, their main advantages and disadvantages. The second stage assessed various aspects such as the use of underground dams in the municipality of Bom Sucesso/PB by applying questionnaires to the municipality's representative authorities as well as to local producers/farmers who use underground dams in their activities, in order to collect socio-economic data, usage, construction site, labour, waterproofing materials used and the use of Amazon wells. It can be seen that the underground dam is efficient for its intended purpose, as it is easy to build and becomes an economical alternative, allowing people to live with drought, ensuring the development of farming activities and the water supply for their homes.

Key words: Living with drought. Rainwater. Groundwater.
Agricultural and pastoral activities. Supply.

SUMMARY

CHAPTER 1

INTRODUCTION

The water crisis that began in 2012 has worsened in recent years and so far continues to cause inconvenience to the residents of the affected regions, especially those with lower purchasing power. Santos et al. (2012) even compared the drought of 2012 to the droughts of 1777-1779 and 1888.

In order to mitigate the situation for residents, especially in the Northeast, which is considered to be the most affected region, drought-fighting measures have emerged, including the construction of dams, cisterns, Operation Pipa, the transposition of the São Francisco River, among others. When it comes to combating drought, the focus is on rainwater harvesting and surface impoundment, which are subject to high rates of evaporation and evapotranspiration. The lack of rainfall harms the recharge of surface reservoirs, which dry up, jeopardising public supply and the development of agricultural and pastoral activities.

In recent years, due to the scarcity of surface water resources, a veritable hunt for groundwater has begun, with Amazon wells being drilled without prior studies anywhere. Groundwater is the result of the infiltration of water from precipitation and the direct recharge of rivers and lakes, and is an indispensable natural resource for life. Of all the water on earth, 97.3 per cent is salt water and 2.5 per cent is fresh water, of which 2.2 per cent is in the form of glaciers. Of the percentage of usable fresh water, 98.5% is groundwater (SANTOS, 2016).

One way of using this groundwater is to build an underground dam, a technology that allows water to accumulate in the soil profile through the use of a waterproofing material, making it possible to grow crops through sub-irrigation, as well as using Amazon wells upstream of the dam to capture the water stored underground and use it for irrigation, human and animal consumption.

Underground dams have been gaining prominence and being propagated because they are a technique that adapts to the conditions of arid and semi-arid regions, helping residents to live with the effects of drought. The use of this rainwater harvesting technology in the municipality of Bom Sucesso/PB began in 2014, as an initiative of the Department of Agriculture, Environment and Development of the Municipality of Bom Sucesso/PB

(SAMAD/PMBS/PB) and has enabled small producers to survive the drought.

CHAPTER 2

OBJECTIVES

2.1 GENERAL OBJECTIVE

The aim is to present the underground dam as an alternative for people to live with the effects of drought, determining its efficiency and the use of this technology in the municipality of Bom Sucesso/PB.

2.2 SPECIFIC OBJECTIVES

- Present the types of underground dam, their purpose and the techniques involved in their construction.
- Carry out a survey of the number of dams built in recent years in the municipality of Bom Sucesso/PB.
- Evaluate the aspects and techniques applied in construction, as well as the purposes for which they are intended in the municipality.
- Present the advantages and disadvantages of this method, and the ways of extracting the water stored in the underground layers of dams.

CHAPTER 3

LITERATURE REVIEW

3.1 DROUGHT: HISTORY, DEFINITION, CAUSES AND CONSEQUENCES

Drought is an ancient phenomenon in northeastern Brazil, with historical references dating back to the time of Portuguese colonisation, as stated by Souza (1979) apud Campos and Studart (2001): "there was a great drought and sterility in the province (Pernambuco) and about four to five thousand Indians came down from the hinterland, occurring to the whites". We realise from the story that in the north-east of Brazil, called "sertão" by the author, there was a great "drought and sterility", in other words, a lack of water resources and a consequent lack of food from these resources, which resulted in the migration of Indians. This is the first historical overview of the drought in north-eastern Brazil and its consequences for the region's inhabitants. According to Campos and Studart (2001), the periodic droughts and adverse conditions in the northeast of Brazil were factors that delayed the beginning of the Portuguese occupation of the region.

In the years 1777 to 1779 there was a great dry spell, responsible for the deaths of many northeasterners. Approximately a century later, in 1888, another drought hit northeastern Brazil, known as "the drought of the three eights". (CAMPOS; STUDART, 2001). We can see, then, that this phenomenon is periodic and has disastrous effects on the inhabitants of these regions, especially those with lower purchasing power.

There are many definitions of drought, and they vary according to the perspective of those who observe it and according to their reality, so there is no definition that is commonly accepted by academics and the general population (SANTOS et al., 2012).

Campos and Studart (2001) define some of the most common types of drought, namely: climatological drought, which occurs due to a lack of total rainfall and has natural causes; edaphic drought, which is characterised by the irregular distribution of rainfall, resulting in severe impacts for the northeast, including economic losses, hunger, misery and migration, which characterise social drought; and finally, hydrological drought, which is characterised by a lack of water in reservoirs, generating impacts on supply. According to

Campos and Studart (2001), "Mitigating hydrological drought depends on efficient water management".

In addition to natural factors (climatic and physical), some authors believe that historical and political factors are closely related to the consequences of droughts. Campos and Studart (2001) state that in the colonial period, in the middle of the 18th and 19th centuries, there was a period without severe droughts, which resulted in an increase in activities such as agriculture and livestock, and consequent population growth. However, constructions aimed at damming water did not keep up with this growth, resulting in a vulnerable population, dependent on small dams and fleeting waters from small surface reservoirs.

Political factors are attributed to the worsening socio-economic situation of the population affected by drought, where some authors use the expression drought industry when they say that funds intended for irrigation projects and water distribution are diverted, failing to benefit the needy population, further aggravating the harsh situation of the residents of these regions affected by droughts (PENA, 2016).

Currently, the drought continues to cause social upheaval in the north-eastern region of the country, damaging supplies, agriculture and livestock farming, which are the main sources of income, especially for small producers.

The state of Paraíba, located in the semi-arid region of northeastern Brazil and included in the so-called drought polygon, has been in a state of emergency since May 2012, through Decree No. 32,984 of 28 May 2012, issued by the state executive; this decree is renewed every six months due to the prolongation of the drought, according to G1 PB (2016).

The municipality of Bom Sucesso/PB has been in an emergency situation since 2012 until this year, and is among the 170 municipalities in Paraíba mentioned in decree No. 32,984. Table 1 shows the rainfall figures for the municipality of Bom Sucesso between 2011 and 2015 according to the Paraíba State Water Management Executive Agency (AESA).

Table 1: Rainfall indices for the municipality of Bom Sucesso between 2011 and 2015

Year	Volume (mm)
2011	776,0
2012	291,3
2013	681,8

2014	703,3
2015	308,1

Source: AESA (2016)

In the midst of this drought scenario, and the consequences caused by this phenomenon, actions to combat it emerge, most of which are political in nature. These include the construction of dams to store water for later use in times of drought, as well as the construction of dams, barreiros, Amazon wells, cacimbas, and more recently, studies aimed at capturing groundwater, since surface water is running out and surface reservoirs need to be replenished by rainfall. Among the means of capturing groundwater are artesian wells and underground dams.

3.2 UNDERGROUND DAMS : A BIT OF HISTORY

The underground dam is one of the alternatives for storing water, in this case underground, and is an alternative for situations of water scarcity because it is a low-cost and easy-to-install technology. It is therefore being used mainly by small producers who don't have a lot of financial resources.

In 1895, California (USA) carried out the first studies into the utilisation of groundwater (TIGRE, 1949). Underground dams for agricultural purposes were built by French hydrogeologists in the Sahara (IPT, 1981). In north-eastern Brazil, specifically in the city of Mossoró/RN, the first underground dams were built by the Inspetoria de Obras Contra a Seca in 1935, with the aim of damming the underground flow of intermittent rivers in the region (IPT, 1981).

According to Benvenuto and Polla (1982), the international dissemination of this rainwater harvesting technology was the responsibility of the United Nations Educational, Scientific and Cultural Organisation (UNESCO), through the Major Project for Arid Zones, in 1951, stating that this technique is suitable for the climatic conditions of the northeastern region. Other authors (COSTA, 1997; SILVA, 1998; SILVA; REGO NETO, 1992) have presented a history of the construction of underground dams in the states of Paraíba in 1919, Rio Grande do Norte in 1920 and Ceará in 1965.

Silva and Porto (1982 apud BRITO et al. 1999) stated that studies carried out in arid and semi-arid regions around the world emphasise the need to store water, especially underground, in order to meet the need for water in rural areas. Therefore, the storage of water underground associated with methods of extracting this water would benefit residents of arid and semi-arid regions, in order to meet human and animal consumption as well as its use in agriculture.

The historical evolution of this rainwater harvesting technology can be seen, from its use abroad to its development in our country. Other studies have been developed in the form of monographs, books and articles, which are of great relevance to the subject. In 1981, the Brazilian Agricultural Research Corporation - Agricultural Research Centre for the Semi-Arid Tropics (EMBRAPA - CPATSA) carried out studies aimed at assessing the performance and appropriate techniques for managing soil and water in underground dams on natural drainage lines, and was one of the pioneers in the use of plastic sheeting as a waterproofing material (BRITO et al., 1989).

3.3 UNDERGROUND DAM

An underground dam is any structure that aims to block the underground flow of a pre-existing aquifer or one created concomitantly with the construction of an impermeable barrier (SANTOS; FRANGIPANI, 1978).

According to Silva et al. (2001) the purpose of an underground dam is to:

> to stop (intercept) rainwater flowing over the surface and into the ground (surface and underground water flow) by means of a wall (impermeable septum) built across the direction of the water. The water from the rain infiltrates slowly, creating and/or raising the water table, which will later be used by plants. This dam stores water in the soil with minimal loss of moisture (slow evaporation), keeping the soil moist for a longer period of time.

Monteiro (1984), Santos and Frangipani (1978) and Silva and Rego Neto (1992) point to two types of dam: underground or submersible, and submerged. These two types of dams have the same objective, but the first has an impermeable septum from the bedrock to a small height above ground level, which can vary between 0.50m and 0.70m, and a small lake can form upstream, while in the second, the impermeable wall is completely submerged in the ground and the captured water is stored in the soil profile. Figure 1 illustrates two models of underground dam: A shows an underground dam using masonry as the waterproofing material, and B shows one using plastic sheeting.

A - Alvenaria

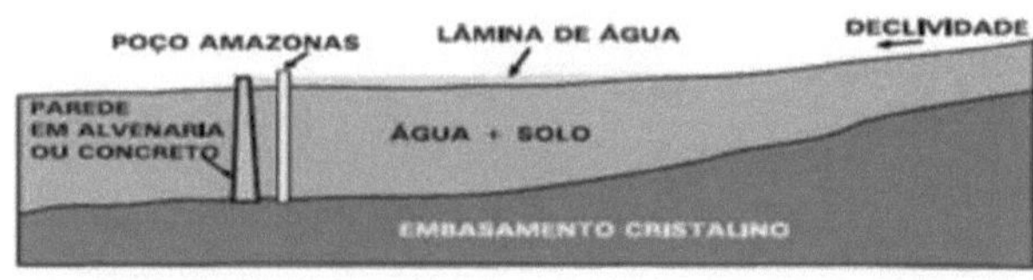

B - Lona plástica

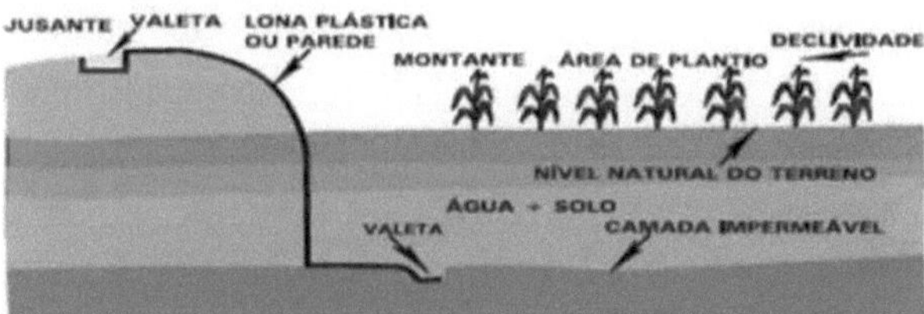

Figura 1: Underground Dams
Source: adapted from BRITO et al. (1999).

According to Costa (2014), an underground dam is a barrier in the alluvial deposit of a river or stream to stop the underground flow when the surface runoff stops. However, Nascimento, Azevedo and Farias (2008) state that in areas with sufficient slope to generate rainwater accumulation, underground dams can be installed for the same purpose, thus taking advantage of the terrain's natural drainage lines, although for this type of dam, the water will only be recharged by rainwater.

It can therefore be concluded that an underground dam is a technology for damming underground water by installing a waterproofing wall, built across the flow of water, which can be in riverbeds and streams or in areas that are steep enough to generate an accumulation of water, allowing the upstream terrain to acquire characteristics such as moist soil, elevation or creation of a water table and low rates of direct evaporation. To be built properly, certain steps must be followed, such as the choice of location, the type of waterproofing material used and the type of dam that will be built.

Over the years, studies involving underground dams have allowed this technique to be improved and some researchers have developed their own construction methodologies, including the CPATSA/ EMBRAPA, COSTA & MELO and CAATINGA models.

1.3.1 Components of an underground dam

The main components of an underground dam according to Brito et al. (1999) are:

- Catchment area (Ac): "is the area represented by a hydrographic basin, formed by the water dividers: topographic and phreatic." (BRITO et al., 1999). The water captured in rainfall flows through the terrain and gradually infiltrates and is stored in the voids in the subsoil. As well as protecting water from evaporation, it protects the soil by reducing erosion.
- Planting area (Ap): this is the catchment area of the underground dam. Rainfall silts up the area upstream of the underground dam, making the soil fertile and suitable for farming. Reservoirs are built in this area, such as the Amazon well, the purpose of which is to capture the water stored underground, as well as to monitor the quality of this water.
- Underground dam wall (Pa): also known as an impermeable septum, its purpose is to stop the underground flow of water. It can be made of different materials, depending on the type of dam you are building. Among the most commonly used are compacted materials (clay, mud), masonry, concrete, polythene or PVC plastic sheeting, the latter being the most widely used, mainly due to its low cost.
- Gully: the purpose of this is to drain the surplus water that accumulates in the catchment/planting area.

1.3.2 Types of underground dams

According to Costa (2002), the types of underground dams are classified based on constructive and functional aspects, as well as by water collection. According to Costa (2002), dams are classified into three types according to water collection:

- Catchment dams: those that dam the underground flow in an attempt to raise the piezometric level of the water table, making it possible to collect the water that has accumulated there.
- Sand and silt accumulation dams: these dam the underground flow, but do not capture the stored water. Upstream of the dam, a small lake is formed which allows agricultural crops to grow.
- Catchment - accumulation dam: this is where the two models mentioned above come together.

According to Costa (1984), underground dams are also classified according to the materials used and the purpose for which they are intended:

- Semi-impermeable dam;
- Watertight dam with clay core;
- Waterproof masonry dam
- Watertight dam with piles.

According to the material used in the waterproofing septum we have:

- Stone or brick masonry;
- Compacted clay;
- Plastic sheeting;
- Metal, wood or concrete piles.

The semi-impermeable dam is used where there is a risk of soil salinisation, so it doesn't retain the water completely, but it can slow down its leaching. Various materials can be used in its construction as long as they allow the water to percolate.

Costa (1997) and Oliveira (2001) present three main models for the watertight barrier, which are presented below in increasing order of complexity and construction costs:

- CAATINGA model: a model proposed by the non-governmental organisation CAATINGA, initially used in Ouricuri/PE. Simple to install, it is made by digging a linear trench and using the material taken from the excavation itself as a waterproofing material, which is then compacted by animals and then rock-filled over the dam. Amazon wells are not used in this model, making it impossible to remove the water stored underground. The stored water is used by the plants through sub-irrigation. Its main advantages include its low cost and easy construction, the use of local labour and no restrictions on its use. The main disadvantages include the lower accumulation of water, which is only used for sub-irrigation, and the fact that the accumulated water cannot be controlled and monitored, which can lead to soil salinisation. Figure 2 illustrates this model of underground dam.

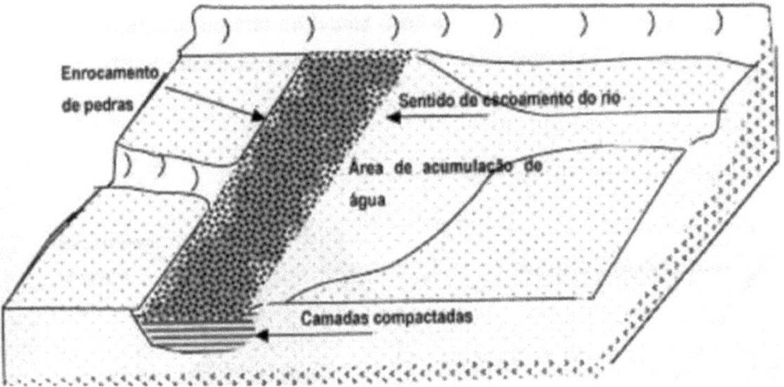

Figura 2: CAATINGA model underground dam Source: COSTA (1997; apud COSTA, 2002).

- COSTA & MELO Model: Developed in the early 1980s by researchers Waldir D. Costa and Pedro G. de Melo, hence the name COSTA & MELO, after the names of its creators. This model has been expanded, modified and adapted to local conditions, and is one of the most widespread. Figure 3 shows a model of this type of dam. It consists of an excavation perpendicular to the direction of the flow, using plastic sheeting as a waterproofing material and a rockfill on top of the dam. In this model, one or more Amazon wells are installed next to the impermeable septum and upstream of the dam, in addition to the use of a piezometer along the hydraulic basin. The advantages include speed and low construction costs, the possibility of collecting and monitoring the stored water through the Amazon wells, which can prevent soil salinisation, and the use of local labour. The disadvantages include the higher cost of the CAATINGA model and the fact that its application depends on specific natural conditions.

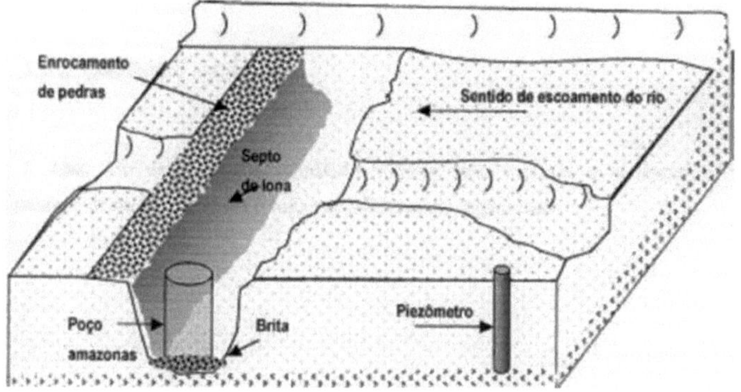

Figura 3: COSTA & MELO model underground dam Source: CIRILO et al. (1998; apud COSTA, 2002).

- CPATSA/EMBRAPA model: Developed in Petrolina/PE in the 1980s by CPATSA/EMBRAPA researchers for the purpose of collecting and storing water and developing subsistence agriculture, it is suitable for medium and large rivers. The excavation is made in an arch, approximately 100 metres long, after which a waterproofing material is used on the wall. Next to the dam wall, a pipe is laid to lead the water stored underground to a cistern downstream of the wall. After the wall has been earthed, a sangradouro is built, which can be made of masonry and cement, or concrete. The main advantages are: greater accumulation of water and separation of the water according to the purpose for which it is to be used.

Among the disadvantages are that this type of dam requires specialised technicians to build; it is more expensive than the other models already mentioned, in addition to the longer construction time; and finally, it does not allow for salinisation control and water level monitoring. Figure 4 shows this type of dam.

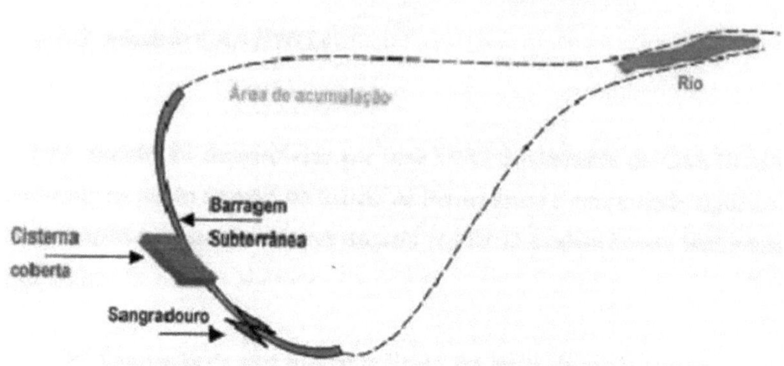

Figura 4: Underground dam model CPATSA/EMBRAPA

Source: COSTA (1997 apud COSTA, 2002).

3.4 LOCATION AND CONSTRUCTION TECHNIQUES

When constructing an underground dam, a number of factors must be taken into consideration in order to ensure a safe construction process that will guarantee the proper functioning of the underground dam. These factors include the choice of installation site, which as mentioned above can be in the bed of a river/stream or in a place with sufficient slope to store rainwater; average rainfall in the region; granulometry of the soils in the selected area; water quality, especially in terms of salinity; depth of the impermeable layer;

slope; storage capacity; distance from the source, among others (BRITO et al., 1999; NASCIMENTO; AZEVEDO; FARIAS, 2008).

The first step is to select the area in which the underground dam is to be installed. According to Brito et al. (1999), after selecting the area it is recommended that a planialtimetric survey be carried out, in 20 x 20 metre squares, in order to better define the location of the dam components. In order to confirm the excavation site for the dam shaft, drilling studies should be carried out. With regard to these drilling studies, Nascimento, Azevedo and Farias (2008) state:

> The purpose of the borehole is to determine the thickness of the alluvial deposit, the morphology of the base of this deposit, the type of soil that

constitutes it, its porosity, permeability coefficient, the water table level and the soil profile at the location of the axis chosen for the dam. It is assumed that the water will be deposited in the soil voids, the accumulated volume is equal to the useful volume of the alluvial deposit multiplied by the porosity.

After the studies, the granulometry of the material present in the alluvium should be analysed, as this will indicate the water storage capacity and the care that should be taken when excavating. The presence of clay material in the soil should be noted, as this can make it difficult to remove the water. In addition, choosing sites with a smaller width will enable greater savings in materials (NASCIMENTO; AZEVEDO; FARIAS, 2008).

After analysing the location parameters, the construction of the underground dam begins. Costa (2014) defines these steps as: digging the trench, installing the waterproof septum, building the Amazon well (for models that use this technology) and then closing the trench. Of the steps mentioned, only the digging of the trench, installation of the waterproof septum and closing the trench are common to all three models mentioned in this work. Other steps used in construction are inherent to the model chosen, including the Amazon well for the Costa & Melo model and the cistern with filtration system for the CPATSA/EMBRAPA model. It's important to note that some owners don't follow one model to the letter, but can mix features from different models, adapting the technology to the reality of their terrain.

The trench can be dug manually or mechanised, always perpendicular to the course of the site where the underground dam is to be installed (whether it be a river/stream or in areas with natural drainage lines). This stage is recommended during the driest periods, when surface runoff ceases and underground runoff is low. The opening of the ditch, also called a trench, must reach the rocky or impermeable material. Figure 5 shows the mechanised excavation of the wall of an underground dam. Figure 6 shows the opening of the wall of an underground dam that intersects the bed of the Bom Sucesso stream.

Figura 5: Opening the trench using a backhoe Source: SAMAD/PMBS/PB archive (2014).

Figura 6: Opening of the ditch intercepting the bed of the Bom Sucesso stream Source: SAMAD/PMBS/PB archive (2014).

Manual excavation is the simplest and least expensive process, using simple equipment that is present on properties, but it does require more time and labour. Mechanical excavation uses machines such as bulldozers and backhoes, the latter being more suitable when the water level is high (COSTA, 2014). Mechanised excavation using backhoes is more common, as it is quicker and more practical.

Some care must be taken at this stage, because depending on the soil and the presence of water, collapses can occur. If there is a risk of this, it is necessary to use trenches to shore up the wall, a process that must be monitored by a specialised technician. The material removed from the excavation must be placed upstream at a safe distance to prevent the material from falling into the open trench, especially when the waterproofing material is plastic sheeting, avoiding inconvenience when installing it and making it easier to drag the material when closing the trench. The open trench must be free of roots and stones.

The waterproof septum is then installed. Various materials can be used for this stage. Brito et al. (1999); Costa (2014) and Nascimento, Azevedo and Farias (2008) present the following materials: compacted material (clay), masonry walls, stones, concrete and plastic sheeting, the latter being the fastest, cheapest and easiest to install.

When built with clay, the ditch must be cleaned, roots and water removed, and then uniform layers of approximately 10 cm must be laid, which after being compacted will be around 5 cm, with the remaining layers being larger, not exceeding 30 cm when compacted with a roller. The masonry wall is more costly and time-consuming, as well as the care that must be taken when building it. This type of wall should be built from well-fired, salt-free bricks or stones, provided the area has enough stones for the construction; it should be raised level, plastered and waterproofed to prevent infiltration. The space between the wall and upstream must be filled with the material removed during excavation. This type of wall requires specialised labour to prevent the dam from malfunctioning and causing damage. For concrete construction, the economic viability must be analysed, as a company specialised in carrying out this type of work is needed, which generally comes at a high cost (BRITO et al., 1999; NASCIMENTO; AZEVEDO; FARIAS, 2008).

The use of plastic sheeting as a waterproofing material was studied in 1989 by EMBRAPA/CPATSA in the model that bears the same name (BRITO et al., 1989). It is a more economically viable material, and the most widespread, especially after the emergence of the Costa & Melo model.

The plastic sheeting is applied to the entire wall at the back of the drain. The site should be cleaned, removing materials that could puncture the sheeting, such as branches, roots and stones. Nascimento, Azevedo and Farias (2008) recommend that 0.5 m of the tarpaulin should rest on the bottom of the trench and 1 m on the downstream side of the land, as shown in Figure 7. Some precautions should be taken when using plastic sheeting as a waterproofing material, such as: wind speed; the presence of perforating material that could tear the sheeting; not tensing it and avoiding times of high temperature, as this can cause the sheeting to expand. The tarpaulin should be attached to the sides with the same mortar used to plaster the wall downstream. Figure 8 shows the installation of the plastic sheeting on an underground dam built in the district of Serrinha, a rural area in the municipality of Bom Sucesso/PB.

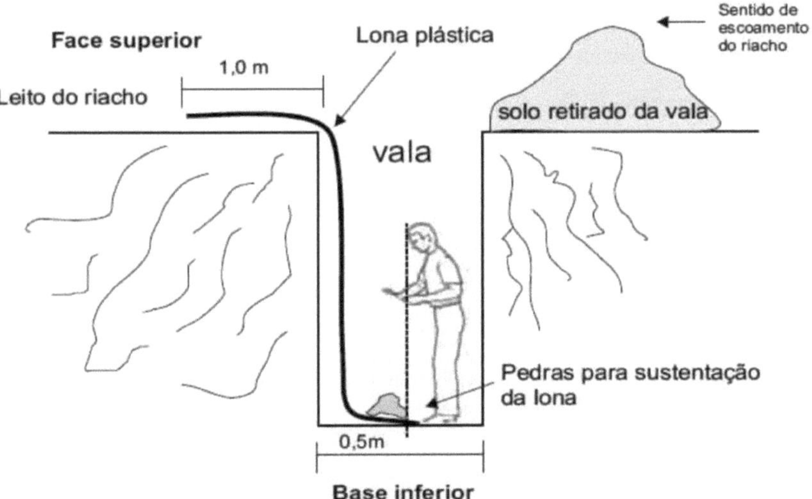

Figura 7: Installing the plastic sheeting

Source: NASCIMENTO; AZEVEDO; FARIAS (2008)

Figura 8: Installation of plastic sheeting on an underground dam built in Bom Sucesso/PB

Source: SAMAD/PMBS/PB archive (2014).

The material removed in the excavation is then pushed in to fill the space between the upstream wall and the material used as a waterproofing, which can be manual or mechanised depending on the process used in the excavation, this stage is called filling the trench. Figure 9 shows this process.

Figura 9: Closing the trench

Source: SAMAD/PMBS/PB archive (2014).

9.4.1 Location and excavation of the Amazonas well

The use of an Amazon well is linked to the type of underground dam you want to build. The COSTA & MELO model was the first to introduce the use of the well, but the idea has been incorporated into the construction of underground dams, even if they are not of the aforementioned model. The Amazon well is installed upstream of the impermeable septum and has a number of functions. Amongst these, the most commonly cited is the abstraction of water retained in the dam for crop irrigation, human and animal consumption. However, the Amazon well is an important tool, as it allows the water to be monitored in order to study its characteristics in terms of salinisation, and this withdrawal of water allows new water from precipitation to infiltrate and thus renew the water resource present there, avoiding salinisation of the soil.

As with the construction of an underground dam, the Amazon well can be dug manually or mechanised, always upstream of the waterproof septum. The COSTA & MELO model recommends locating the well close to the waterproof wall, taking advantage of the trench already excavated to place the waterproofing material (COSTA 2014). Nascimento, Azevedo and Farias (2008) recommend locating the well at least 2 metres upstream from the underground dam wall. The material used can be brick masonry or pre-moulded concrete rings.

When the excavation is done by hand, the precast concrete ring serves as a mould and is installed as the excavation progresses until it reaches sound rock or impermeable material. When construction is mechanised, a rectangle is opened with a diameter equal to the diameter of the precast ring plus 1 metre (NASCIMENTO; AZEVEDO; FARIAS, 2008).

To install the rings with the aid of a machine, two workers are needed to tie the ring to the machine, which will take the ring to the excavation site. When installing the rings by hand, at least six workers are needed to lower one ring over the other. Safety must come first, avoiding sudden movements and people entering the excavation site in order to avoid accidents.

The last ring should be grouted with cement and covered with a lid to protect the water in it. Once the rings have been installed and the well built, the lateral spaces should be filled in, preferably with sandy soil or with the material removed during excavation (NASCIMENTO; AZEVEDO; FARIAS, 2008).

9.4.2 Excavation and location of BAPUCOSA

Some authors recommend the use of successive underground dams to improve performance, providing a greater accumulation of rainwater. Other techniques are also used for this purpose, such as rockfill, used in the COSTA & MELO model. A mechanism to support underground dams is the tyre dam used to contain soil and water (BAPUCOSA).

When torrential downpours occur, the soil is carried away and the water takes longer to infiltrate, which is not enough to generate the accumulation of water inside underground dams, especially in clay soils. The BAPUCOSA consists of a wall made of tyres with the aim of retaining soil and water, thus providing greater infiltration and consequent storage in the dam, as well as increasing the amount of organic matter upstream of the dam after the sedimentation of soil particles suspended in the water.

According to Azevedo et al (2010), the BAPUCOSA is built approximately one metre downstream from the dam, in the shape of an arch. The topsoil is removed so that the excavation is lower than the height of the tyre, allowing for greater stability and preventing it from breaking underneath. It is recommended to use the same type of tyre to build the layers, so the tyres are laid on top of each other so that the centre of the tyre in the previous layer coincides with the edges of the tyres in the upper layer. A wall with up to four layers is recommended; beyond that, the structure will require more secure tying. After installing the layers, stone should be placed inside the tyres to provide greater resistance to flooding.

For tyre tying, rebar with a gauge of ^ inch is recommended, with the length varying according to the depth of the waterproof layer at the site where the dam is being built. With each layer built, a rod is placed next to the upstream face of each tyre, and a sledgehammer is used to penetrate the rod so that approximately 40 cm is left over to tie the tyre down. This procedure fixes the tyre in the same alignment as the layer below (AZEVEDO et al, 2010).

9.4.3 Underground dam management and the risk of soil salinisation

A major challenge and one that some authors also point to as a disadvantage in the use of underground dams is the process of soil salinisation, as it generates an accumulation of salts in the soil, making it infertile and unsuitable for agricultural use, and could culminate in the process of desertification.

The water present on the surface or underground has dissolved salts in its composition and in arid and semi-arid regions the concentration of salts is higher due to dry periods (NASCIMENTO; AZEVEDO; FARIAS, 2008). The process of salinisation is linked to physical and chemical factors that contribute to an increase in the concentration of salts in the soil. These factors include evapotranspiration, where water is evaporated but the salts are retained in the soil, increasing their concentration. Another factor that contributes to this process is the low levels of precipitation and the irregular distribution of rainfall, making it impossible to wash the soil to remove the salts.

Anthropogenic actions also contribute to the formation of saline soil, as the improper use of pesticides can aggravate the salinisation process. If the water accumulated in the dam is intended for human consumption, it is not recommended to use pesticides on crops upstream of the dam to avoid contaminating the water.

Underground dams are subject to the process of soil salinisation, as the water retained can be evaporated in such a way that the concentration of salts makes the soil unsuitable for cultivation, which is one of the objectives of the underground dam.

With a view to proper management in order to avoid salinisation, some authors have proposed measures to avoid and/or mitigate the effects of salinisation. The CPATSA/EMBRAPA model proposes the use of a discharge pipe, passing through the impermeable septum to the downstream end of the dam, allowing the water to wash away the soil and carry away the salts present. The Costa & Melo model proposes the use of Amazon wells upstream of the dam for the same purpose as the discharge pipe used in the

CPATSA/EMBRAPA model. In this way, the water present in the soil is removed, allowing new water from precipitation to infiltrate and wash the soil profile.

3.5 CALCULATING THE VOLUME OF AN UNDERGROUND DAM

The low accumulation volume of an underground dam is seen as a disadvantage of using this technology, but is it possible to estimate a value for the accumulation volume? The specialised literature confirms that it is through parameters obtained during construction and through laboratory studies.

Nascimento, Azevedo and Farias (2008) state that the amount of water in an underground dam depends on characteristics such as density, porosity and texture, and that knowledge of the amount of water available is important for planning the use of the available water resource. Also according to Nascimento, Azevedo and Farias (2008), the volume of an underground dam can be calculated by taking into account factors such as: soil moisture on a volume basis, the depth of the layers and the surface area. The volume calculation uses Equation 1:

$$V_n = \frac{U_{VN} * Z_N * A_S}{100} \quad \text{Eq. (1)}$$

In which:

-Uvn represents soil moisture on a volume basis cm /water/cm /soil;[33]

-Zn represents the depth of the layers in cm;

-A_s represents the surface area of the dam in m^2

-N represents the number of layers

For a better understanding, it is necessary to divide the underground dam into layers as shown in Figure 10.

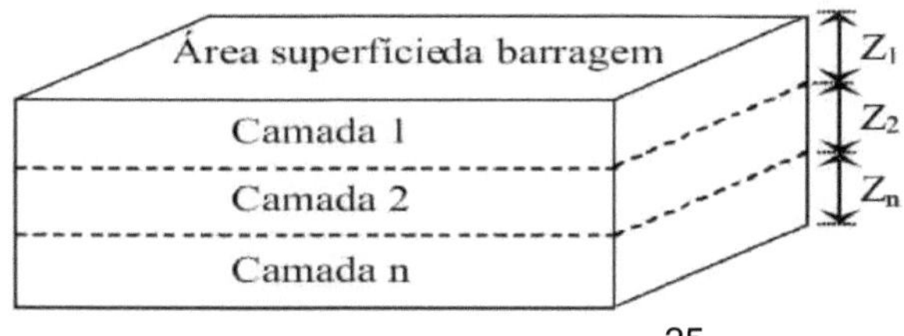

Figura 10: Dividing the underground dam into layers

Source: NASCIMENTO; AZEVEDO; FARIAS (2008)

According to Costa et al. (1998), another way of calculating the volume of an underground dam takes into account the width of the valley, the extent of the accumulation area, the thickness of the alluvial deposit and the porosity of the alluvium, and Equation 2 can be used:

$$V = L * Ex * Es * \mu \quad \text{Eq. (2)}$$

In which:

- V represents the volume in m^3
- L - valley width
- Ex - extension of the accumulation area
- Es - thickness of alluvial deposit
- μ - alluvial porosity

CHAPTER 4

METHODOLOGY

4.1 FIRST STAGE

The first stage consisted of bibliographical research, reviewing published works, books and dissertations, seeking to analyse, through the results achieved by the authors, whether underground dam technology is efficient in dealing with the effects of drought; the results in relation to its use in different locations and finally to present the advantages and disadvantages of an underground dam.

4.2 SECOND STAGE

The second stage of the work was an analysis of the use of underground dams in the municipality of Bom Sucesso/PB, through the application of an interview and questionnaires, which are presented in the annexes to this work, with the aim of evaluating the results of using the technology in the municipality.

The interview was directed at the municipal secretary of Agriculture, Livestock and Supply, in order to gather data on the programme to build underground dams in the municipality of Bom Sucesso/PB, a programme initiated by the municipal government, through the Municipal Secretariat of Agriculture, Environment and Development (SAMAD).

The questionnaires were addressed to owners of underground dams, both those covered by the public authorities and those who had built underground dams on a private initiative, with the aim of collecting data such as: the socio-economic status of the target public and their use of the dam, the use of the Amazon well, the income obtained from the use of the dams, the results presented, etc.

4.2.2 Characterisation of the second stage study area: The municipality of Bom Sucesso/PB

4.2.2.1 Location and access

The municipality of Bom Sucesso is located in the western region of the Paraiba hinterland, in the Catolé do Rocha microregion. It is bordered to the north by the towns of Alexandria/RN and Brejo dos Santos/PB, to the east by Brejo dos Santos/PB and Jericó/PB, to the south by Lagoa/PB and Santa Cruz/PB, and to the west by Alexandria/RN and Santa Cruz/PB. Figure 11 shows an aerial view of the municipality of Bom Sucesso/PB

Figure 11: Location of the municipality of Bom Sucesso/PB

Source: Google Earth (2016)

4.2.2.2 Socio-economic aspects

Created by law 3,049 of 17 June 1963 and installed on 02 August 1963, the municipality of Bom Sucesso/PB has an area of 184.102 Km^2 and a population of 5,035 inhabitants according to the last census of 2010, of which 2,030 (40.32%) live in the urban area and 3,005 (59.68%) in the rural area, thus having a population density of 27.35 inhabitants/Km^2 , according to the IBGE 2016 website.

The municipality of Bom Sucesso has a Gross Domestic Product (GDP) of R$20,325,000.00. The breakdown of GDP can be seen in Graph 1.

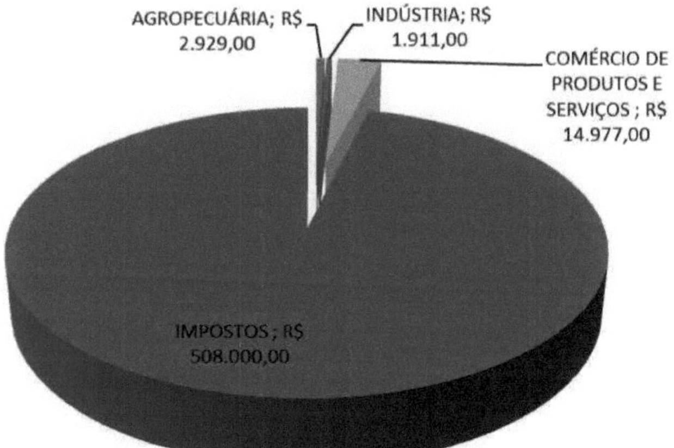

Graph 1: Breakdown of GDP in the municipality of Bom Sucesso/PB

Source: Prepared by the authors (2016)

4.2.2.3 Physiographic aspects

Located in the drought polygon, the municipality of Bom Sucesso has a hot and humid climate with autumn and winter rains. According to the zoning of the state of Paraíba into bioclimatic regions, the municipality has a tropical climate, which is hot and has mild droughts, ranging from 7 to 8 dry months. The average rainfall is 1,000 mm, the vegetation is of the Caatinga type, and the average annual temperature is between 27°C and 28°C.

CHAPTER 5

DEVELOPMENT

5.1 THE SUCCESS AND/OR FAILURE OF UNDERGROUND DAMS IN DIFFERENT PARTS OF THE COUNTRY.

The construction of underground dams has been widespread and used in our country, mainly in regions affected by droughts, such as the northeast, but other regions of the country use this technology, such as the southeast, in the states of Rio de Janeiro and São Paulo.

Despite all the success of underground dams, certain precautions must be taken when using them, especially with regard to the risk of soil salinisation. To this end, location and construction techniques, together with soil and water analyses, and proper management of the use of underground dams are necessary. Many studies on this rainwater harvesting technology analyse the success or failure of its use.

The use of underground dams is widespread in several northeastern states, including Paraíba, Rio Grande do Norte, Ceará and Pernambuco. Experiences in these states show satisfactory results and they are used for a wide variety of purposes, such as irrigation, animal watering, human consumption, family farming, rainfed agriculture, forage, etc.

Most of this work has focused on analysing underground dams in the state of Pernambuco, which concentrates a good number of this type of construction. According to Costa et al (1998), in 1998, at the initiative of the state government and the Departments of Science and Technology, the Environment and Agriculture, the state of Pernambuco had a programme to build 3,000 underground dams throughout the state.

In 1981 EMBRAPA/CPATSA built three underground dams using the model that bears the same name as the company, achieving results such as guaranteeing risk-free agricultural crops in years of irregular rainfall, up to two harvests in regular years; low losses due to evaporation and evapotranspiration; low implementation costs and easy management. During three years of research on caupi, maize and sorghum crops, yields varied from 542 to 1093 kg/ha, 2128 to 4697 kg/ha and 2993 to 4531 kg/ha, respectively (BRITO et al., 1989).

Brito et al (1999) presented results at producer level from an underground dam built in 1994 in the municipality of Alexandria/RN, in an area of 29.1ha, where in 1995 the owner harvested 100 50 kg bags of rice, which corresponds to approximately 2.5 $t.ha^{-1}$, whereas in previous years, in the same area, a maximum of 300 kg was harvested. The author also points out that with the resources provided by the dam it was possible to buy an irrigation kit, using the Amazon well present in the dam to irrigate 1.0 ha downstream of the dam.

According to Costa (1987), in 1985 some underground dams of the COSTA&MELO model were built on a property in the municipality of São Mamede/PB. Some Amazon wells were built upstream of the dam and used to irrigate an area of 40 hectares. The use of the dam allowed the owner to export mangoes to Europe, as well as grow other crops through sub-irrigation, such as corn, beans, grass and fodder crops for cattle breeding. The 4 metre diameter Amazon wells were designed to pump $30m^3$ /h each with a pumping regime of 8 hours/day.

According to Palmier and Carvalho (2003), around 30 per cent of the total area of the state of Minas Gerais is semi-arid. In the 1980s, a number of dams were built in the state, mainly in the Jequitinhonha and São Francisco river basins. With regard to these underground dams, information is scarce, but the author concludes that the use of this technology in the state has not been successful and that the failure was due to a lack of training for both the executors and the users.

Costa et al. (2000) presented in their work an apparent failure in the use of underground dams in the state of Pernambuco, due to a number of factors in their location and construction, including: the construction of the dam at the headwaters of rivers; the reduced thickness of the alluvial fans; clay soil; the existence of surface dams nearby; the presence of rocky sills in riverbeds; salinised water; improper location of the shaft; the absence of rockfill for the COSTA&MELO model; the Amazon well with impermeable pipes and partial waterproofing of the septum.

When built inappropriately, underground dams cause financial losses, potentially ruining the positive image of this technology for living with drought, because, as other studies have shown, the use of dams is efficient and generates returns for the owners in a wide variety of uses.

5.2 ADVANTAGES AND DISADVANTAGES OF USING UNDERGROUND DAMS

One of the advantages of an underground dam is that it allows agricultural activity to flourish even in the midst of scarcity, since it allows the level of groundwater to rise, which can be used for sub-irrigation, where it is taken up by the plants through their roots, or for irrigation, using the Amazon well to take up the water.

The stored water is free from the sun's rays, providing protection against evaporation, which according to Brito et al (1999) can reach an average of 2000 mm per year and according to Costa (2014) can reach 2500 mm/year or 7mm/day, mainly affecting surface reservoirs.

According to Nascimento, Azevedo and Farias (2008), other advantages that can be cited are: no flooding of the land, which can be used for planting; low risk of breakage; ease of repairing the system and, finally, its construction and use has a lower environmental impact compared to surface dams, mainly because the area where the water accumulates is underground. Surface dams flood large areas, destroying fauna and flora and sometimes causing the displacement of villages, towns or cities, causing negative environmental and social impacts.

The cost of an underground dam compared to other rainwater harvesting methods is one of the biggest advantages of this technology, as it eliminates the need for treatment, maintenance, operation, electricity and other costs associated with surface dams. Associated with the low cost is the ease of construction and the use of local labour, generating income for the population and speeding up construction. Dams take an average of one to two days to complete if construction is mechanised, as well as better protection of the water against surface bacterial contamination (BRITO et al., 1999; COSTA et al., 1998; NASCIMENTO; AZEVEDO; FARIAS, 2008).

Among the disadvantages, the biggest concern is the risk of salinisation, which can be caused by evaporation or the use of fertilisers on crops. However, some precautions can be taken to avoid salinisation, such as a preliminary analysis of the soil and water, because if they are already saline, the tendency is to get worse, and it is not advisable to use this land.

The use of an Amazon well removes the water dammed up in the reservoir, allowing new water to seep in, washing the soil and preventing salinisation.

Another disadvantage is the volume of accumulation, which is low compared to surface dams. Brito et al. (1989) and Nascimento, Azevedo and Farias (2008) recommend the use of successive underground dams, forming an integrated system that allows for greater accumulation of water. The use of BAPUCOSA, which obstructs the flow of surface water, allows for greater infiltration, keeping moisture in the soil and allowing the development of annual crops and the maintenance of perennial crops.

5.3 ANALYSING AND INTERPRETING THE RESULTS OBTAINED FROM THE INTERVIEWS AND QUESTIONNAIRES

The municipality of Bom Sucesso/PB has been affected by the effects of the drought since 2012, when it had the lowest rainfall of the last five years, according to AESA, with a volume of 291.3 mm for a municipality that has an annual average of 1000 mm of rainfall.

In the following years, the situation worsened, making public supply difficult and the lives of small producers who are used to planting after the first rains and irrigating with the water that accumulates in Amazon wells or clay pits. Faced with this water shortage, and the consequent worsening of the social drought, the municipality of Bom Sucesso/PB sought to mitigate the situation by developing programmes to combat the drought, in conjunction with the Municipal Secretariats for Agriculture, Livestock and Supply and the Secretariat for Infrastructure. One of the programmes involved rural landowners building underground dams. Information on this programme was obtained through an interview with the municipality's Municipal Secretary for Agriculture, Livestock and Supply.

At the end of 2013, machines arrived in the municipality from the Federal Government's Phase 2 Growth Acceleration Programme (PAC 2), with the municipality receiving a backhoe, a filler, a motor grader, a bucket and a water tanker. With the arrival of the machines, the municipal government began a programme to combat drought, building and restoring small dams, reservoirs, drinking fountains and underground dams. To implement the municipal activities, partnerships were formed between the municipal government, the Technical Assistance and Rural Extension Company of Paraíba (EMATER), the Rural Workers' Union and the farmers/producers.

The idea of underground dams was initially questioned, as it was a previously unknown technology. However, the municipal secretary of agriculture took part in a training course, given by EMATER technicians, which covered knowledge about underground dams, such as location techniques, construction and the use of tarpaulin as a waterproofing material. The knowledge acquired was passed on to the municipal government, which accepted the idea and went ahead with the project to build the dams.

Construction of the dams began in 2014, with the municipality providing the machinery and operators and the landowners paying for the tarpaulin and labour. The underground dams built by the municipality were gradually transformed into small reservoirs by the landowners, leaving a total of 22 underground dams built by the municipality and in full operation. The government spent around R$1,200.00 to R$1,500.00 per dam, while the landowners paid between R$500.00 and R$700.00 per dam built.

With the construction of the underground dams by the municipality and the start of the results, even with the low rainfall, some landowners built underground dams on their own initiative.

The landowners who were included in the municipal executive's dam-building programme and those who built dams on their own initiative were the subject of this study, and answered a questionnaire.

Of the 22 underground dams built by the municipality, twelve were visited, corresponding to 54.54% of the total number of dams. As there is no data on the number of privately built dams, it is not possible to say for sure how many have been built in recent years, but according to popular reports, six have been built. The on-site visit was made to three privately built dams, distributed as follows in the communities of: Oiticica dos Belos, Passagem Molhada and the Bom Sucesso site. The data obtained was analysed and the results are presented below:

As for the breakdown by gender, of the 15 underground dam owners surveyed, thirteen are male, while two are female. Graph 2 shows the percentage breakdown by age group. It can be seen that the largest number of underground dam users are aged over 40. Of the 15 interviewees, twelve were aged over 40, one was aged between 18 and 30 and two were aged between 31 and 40.

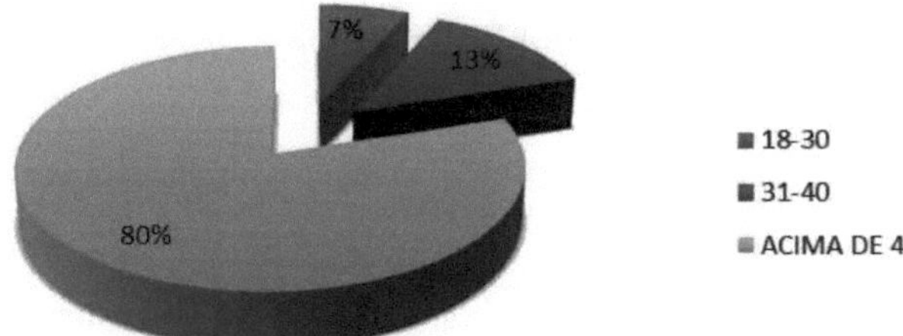

Graph 2: Breakdown by age group

Source: Prepared by the authors (2016)

With regard to the construction method, all 15 owners interviewed said that they used machinery to excavate and then backfill the wall of the underground dam. The dams built by the municipality used backhoes and fillers. The privately built dams used tracked backhoes, given the depth of the solid foundation. For example, the underground dam built in the Passagem Molhada community was 13 metres deep.

With regard to preliminary studies, i.e. prior studies aimed at the best performance and execution in the construction of the underground dam, a number of types of studies were listed, among them: topographical survey aimed at the best place to open the wall; drilling and granulometric analysis to gauge the type of soil where the construction will be carried out and, finally, soil and water analyses aimed at obtaining data on the salt content, thus avoiding the risk of salinisation. The survey showed that none of the interviewees had carried out any prior studies on the construction of underground dams.

With regard to the material used in the waterproofing septum, the use of plastic sheeting was predominant, i.e. 100% of those interviewed used plastic sheeting when building underground dams. This is due to the fact that among the materials that can be used in the waterproof septum, plastic sheeting is the most cost-effective, as it is cheap, easy to install and effectively fulfils its intended purpose. When using plastic sheeting, care must be taken when installing it to avoid perforations in the sheeting and the subsequent escape of trapped water. To do this, the wall must be cleaned to remove any branches, stones or other materials that could damage the plastic sheeting.

All the interviewees used local labour. This is pointed out as one of the advantages of building an underground dam, as it is easy to execute and does not require specialised labour.

As well as saving money on the use of skilled labour, local labour generates income for the municipality.

The location where the underground dam wall will be installed is of fundamental importance, as any mistakes at this stage can jeopardise subsequent stages. The underground dam can intercept riverbeds or streams, and can also be built in areas that are steep enough to accumulate water. Graph 3 shows that only 20 per cent, i.e. three underground dams, were built across river/stream beds, and these were the three dams built by the private sector. The other 80%, corresponding to twelve dams, were built in areas with sufficient slopes to generate water accumulation. It is important to emphasise that both are functioning normally, showing good results, regardless of where the wall is installed.

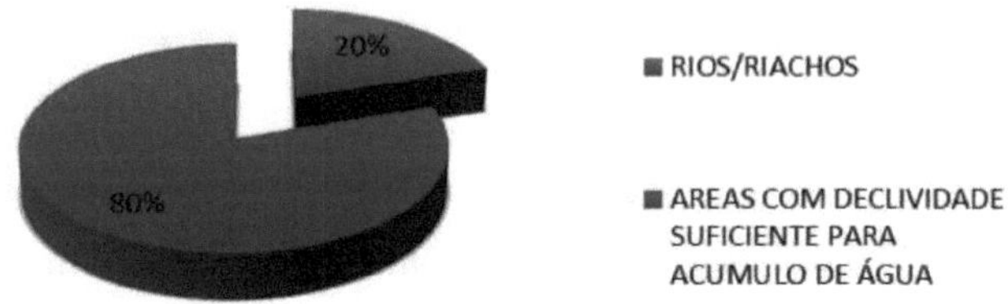

Graph 3: Underground dam wall installation location

Source: Prepared by the authors (2016)

With regard to the purposes for which underground dams are used, some of the most common uses for this technology were considered, among them: agriculture, a source of income for many Bomsucessense families. Maize and beans were grown upstream of the dam, using sub-irrigation and irrigation for the dams that had an Amazon well; in livestock farming, the highlight was cattle and pig farming, where the dam was used for animal watering, using the water obtained from the Amazon well, and for growing fodder crops used to make animal feed.

Graph 4 shows the values, in percentages, for the different uses of the underground dam. There was little use for agriculture and growing fodder crops due to the low rainfall, which made it difficult to collect the water. Livestock and animal watering stand out because, even with little rainfall, the water impounded by the dams allows producers to at least quench their animals' thirst.

Of those interviewed, 9%, or four interviewees, use the dam for human consumption, collecting water from the wells and using it for a wide range of domestic chores; in addition to home consumption, two of the interviewees said that they use the water from the dam to supply their homes in the city, given that the public supply has been suspended. A thousand litres of water are sold for R$15.00.

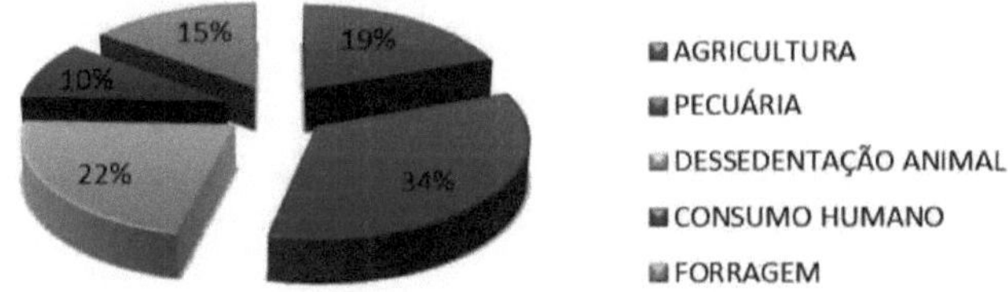

Graph 4: Uses of Underground Dams

Source: Prepared by the authors (2016)

The Amazon well has several applications and is of fundamental importance when used in conjunction with an underground dam. Through the well, water can be collected for irrigation, animal watering and human consumption, as well as allowing the water to be analysed for salinisation risks. According to Graph 5, of the fifteen interviewees, nine have an Amazon well, corresponding to 60% of those interviewed; the other six do not have an Amazon well, but have shown interest in building one, with financial difficulties being the biggest obstacle. Graph 6 shows that of the nine owners who have an Amazon well on their dams, only one built the well after the dam was built, the other eight had already built an Amazon well on their land.

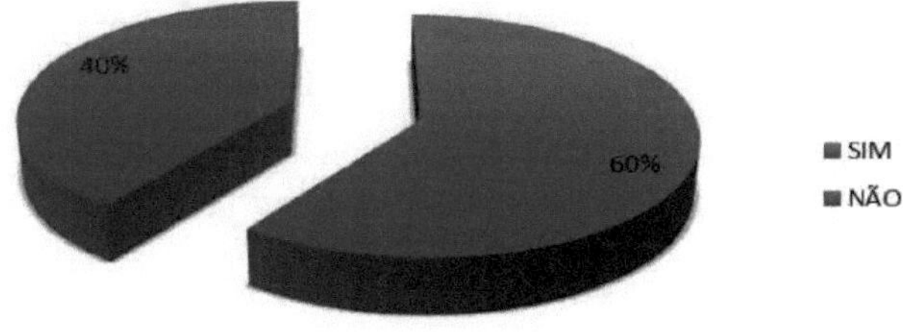

Graph 5: Owners who have an Amazon well next to an underground dam

Source: Prepared by the authors (2016)

Graph 6: Amazon well construction in relation to underground dam construction

Source: Prepared by the authors (2016)

With regard to the model of dam used, it was observed that of the 15 dams visited, all were built using the same model, not following a specific model among the three most commonly used: CAATINGA, COSTA e MELO and CPATSA/EMBRAPA. The construction of the dams consisted only of opening the ditch, laying the tarpaulin and closing the ditch. No other components inherent to the aforementioned models were built, such as Amazon wells, cisterns, stone hauling, etc.

With regard to the results presented by underground dams, the interviewees expressed their opinion that the dams have shown good results, even in the midst of a lack of rainfall to recharge the water in the dams.

Graph 7 shows the results, in percentages, of the degree of user satisfaction with the results presented by the dams. Of the 15 interviewees, thirteen (87 per cent) were satisfied, while 13 per cent (two) were dissatisfied with the results of the dams.

A visit to the site revealed that the dams that didn't live up to users' expectations may have had errors in their construction. When we spoke to the owners who were dissatisfied, they said that the problem was with the closing of the ditch, and that because the ditch wasn't closed properly, the animals trampled on the tarpaulin, causing holes to form and water to leak out.

Among the results presented, the most cited were: an increase in water, both for human and animal use and irrigation; more moisture in the soil; a reduction in feed costs; an increase in profits for those who used the dams to sell water, among others.

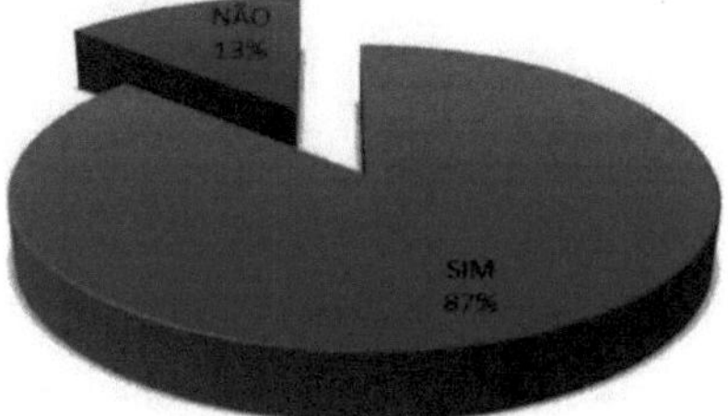

Graph 7: Respondents' level of satisfaction with the use of dams. Source: Prepared by the authors (2016)

CHAPTER 6

FINAL CONSIDERATIONS

The use of underground dams can mitigate the effects of drought, allowing crops to grow and animals to be raised and consequently improving the quality of life of the populations affected by this phenomenon. As it is a low-cost and easy-to-implement alternative, it makes it possible to live with drought, even for the poorest populations, who can get together and build a dam that benefits the whole community, using local labour and sharing the costs.

In relation to the municipality of Bom Sucesso/PB, this study showed that the use of underground dams has improved the economic and social situation of small producers, allowing them to grow crops during the dry season, as well as raise animals and provide water for human consumption. The satisfaction of those who have an underground dam on their property is notorious. As well as enabling the development of agricultural and pastoral activities, the underground dam adds value to the land, generating even more profits for the owner.

However, it's worth pointing out that because they didn't present any previous studies

In terms of soil, water and salinisation, the dams in the municipality of Bom Sucesso/PB run the serious risk of their soils becoming saline, thus damaging their performance. Another aggravating factor is that some dams don't have Amazon wells or any other type of technology that allows the water stored in the soil profile to be removed and then replaced by new rains, which can further accelerate the salinisation process, as they don't allow the new water from the rainfall to wash the soil.

The programme to build underground dams in the municipality of Bom Sucesso/PB is commendable and deserves recognition, although improvements are needed. More investment is needed, from all three spheres - federal, state and municipal - as well as training courses for dam users, so that they are more knowledgeable and can be properly managed, thus preventing a simple, low-cost and efficient technology from becoming inefficient and causing harm to its users.

REFERENCES

AZEVEDO, M. A.; NASCIMENTO, J. W. B.; FURTADO, D. A. **Construction techniques for underground dams, BAPUCOSA and Poços Amazonas**. Revista Educação Agrícola Superior. Brasília, v.25, n.1, p.31-36, 2010.

BENVENUTO, C.; POLLA, C. M. **Geotechnical aspects of the design and construction of underground dams in the northeast.** In: BRAZILIAN CONGRESS OF SOIL MECHANICS AND FOUNDATION ENGINEERING. 7., 1982, Olinda - Recife, PE. ANNALS - Recife. ABMS. V. 2. p. 417-428.

BRITO, L. T. de L.; SILVA, A. de S.; MACIEL, J. L.; MONTEIRO, M. A. R. **Barragem subterrânea I: construção e manejo.** Petrolina, PE: EMBRAPACPATSA, 1989. 38 p. (EMBRAPA-CPATSA. Boletim de Pesquisa, 36).

BRITO, L. T. de L.; SILVA, D. A. da; CAVALCANTI, N. de B.; ANJOS, J. B. dos; RÊGO, M. M. do. **Technological alternative to increase water availability in the semi-arid region**. Revista Brasileira de Engenharia Agrícola e Ambiental, Campina Grande, PB, v. 3, n. 1, p. 111-115, 1999.

CAMPOS, J. N. B.; STUDART, T. N. C. **Droughts in the Brazilian Northeast: Origins, Causes and Solutions**. IN: IV Inter-American Dialogue on Water Management, Foz do Iguaçu, 2001.

COSTA, M. R. **Evaluation of the utilisation potential of reservoirs made up of underground dams in the Brazilian semi-arid region.** 2002. 198 f. Dissertation (Master's Degree) - Postgraduate Programme in Civil Engineering, Department of Civil Engineering, Federal University of Pernambuco, Recife-PE, 2002.

COSTA, W. D. **Alluvial Aquifers as Agricultural Support in the Northeast.** In: BRAZILIAN CONGRESS OF UNDERGROUND WATERS. 3, Fortaleza - CE, Proceedings... Fortaleza, 1984. v.1. p.431-440.

COSTA, W. D.; CIRILO, J. A.; PONTES, M.; MAIA, A. Z.; SOBRINHO, O. P. **Underground Dam: An Efficient Way of Living with Drought**. Available at : <https://aguassubterraneas.abas.org/asubterraneas/article/view/22277/14620> Accessed on: 25 Mar. 2016.

COSTA, W. D. **Underground dams vs. siltation dams: construction aspects and purposes**. Disponível em: <https://aguassubterraneas.abas.org/asubterraneas/article/view/28311/18420> Accessed on: 25 Mar. 2016.

COSTA, W. D. **Manual of underground dams. Basic concepts, location and construction aspects.** Department of Science, Technology and Environment of Pernambuco-PE, 1997.

COSTA, W. D., CIRILO, J. A ABREU, H. F. G. and COSTA, M. R. **The apparent failure of underground dams in Pernambuco**. In: INTEGRATED WORLD CONGRESS ON UNDERGROUND WATER, 1; BRAZILIAN CONGRESS ON UNDERGROUND WATER, 11, 2000, Fortaleza. Proceedings... Fortaleza: ABAS; AHLSUD; IAH, 2000. 1 CD-ROM.

COSTA, W. D. (1987). **Hydrogeological research for the implementation of underground dams in alluvial aquifers**. Proceedings of the 1st Northeast Hydrogeology Symposium, ABAS, Recife-PE, p.13-23.

G1. **Government extends state of emergency in 170 cities in PB**
Available at <http://g1.globo.com/pb/paraiba/noticia/2016/04/govemo-

extends state of emergency in-170-cities-of-pb.html>. Accessed on 25 March 2016.

IPT - INSTITUTO DE PESQUISAS TECNOLÓGICAS, São Paulo - SP. **Survey of the potential for implementing underground dams in the Northeast: the Piranhas-Açu RN and Jaguaribe CE river basins**. São Paulo: IPT, 1981. 56p. il. Report 14887.

MONTEIRO, L. C. **Underground dams: an alternative for water supply in the semi-arid region.** In: BRAZILIAN CONGRESS ON UNDERGROUND WATER, 3, Fortaleza, CE. 1984. **Proceedings...** Fortaleza, ABAS, 1984. v.1. p.421-430.

NASCIMENTO, J. W. B.; AZEVEDO, M. A.; FARIAS, S. A. R. **UNDERGROUND DAMS**. Campina Grande / Pb: Gráfica Agenda, 2008. 96 p.

OLIVEIRA, J. B. **PRODHAM technical operating manual**. Fortaleza: SRH, 2001.

PALMIER, L. R.; CARVALHO, F. R. **Underground Dams: the Experience of the State of Minas Gerais**. In: 4th Brazilian Symposium on Rainwater Harvesting and Management, Juazeiro, BA, 2003.

PENA, R. F. A. **"Why is there drought in the Northeast?"**; *Brasil Escola.* Available at <http://brasilescola.uol.com.br/brasil/por-que-nordeste-seco.htm>. Accessed on 25 March 2016.

SANTOS, A. R.; **WATER IN NATURE AND THE HYDROLOGICAL CYCLE.** Availabl e in <http://mundogeomatica.com.br/CL/ApostilaTeoricaCL/Capitulo14-AguaNaturezaCicloHidrologico.pdf> Accessed 15 May 2016.

SANTOS, E.; MATOS, H.; ALVARENGA, J.; SALES, M. C. L. **The drought in the Northeast in 2012: Report on the drought in the Northeast region and the example of practice of coexistence with the semi-arid in the District of Iguaçu/Canindé/Ce.** Geonorte Magazine, Special Issue 2, V.1, N.5, p 819 - 830, 2012.

SANTOS, J. P. dos; FRANGIPANI, A. **Submerged dams - an alternative for the Brazilian Northeast.** In: BRAZILIAN CONGRESS OF ENGINEERING GEOLOGY, 2, 1978, São Paulo. **Proceedings...** São Paulo: ABGE, 1978. v.1. p.119- 126.

SILVA, D. A.; REGO NETO, J. **Evaluation of submersible dams for agricultural exploitation in the semi-arid region.** In: CONGRESSO NACIONAL DE IRRIGAÇÃO E DRENAGEM, 9, Natal - RN, 1992. **Annals** Natal: ABID, 1992, v.1. p.335-361.

SILVA, F. F. (1998). **Investigation and Modelling of Underground Flow in an Alluvial Aquifer in the Semi-Arid Region of Paraíba**, Hydraulics Laboratory - DEC/CCT/UFPB,

Dissertation, Master's Degree in Water Resources Engineering, Campina Grande-PB, 109 p.

SILVA, M. S. L.; ANJOS, J. B.; BRITO, L. T. de L.; SILVA, A. de S. S.; PORTO, E. R.; HONÓRIO, A. P. M. **Underground dam**. Petrolina: Embrapa Semi-Arido, 2001 (Embrapa Semi-Arido. Technical Instructions, 49).

TIGRE, C. B. **Underground and submerged dams as a fast and economical means of storing water.** Anuário do Instituto do Nordeste, Fortaleza, CE, p.13-29, 1949.

APPENDIX A - Interview with the Municipal Secretary for Agriculture, Livestock and Supply.

1. Where did the idea of building underground dams in the municipality's rural communities come from?
2. What is the municipality's objective?
3. Were there any partnerships? ( ) YES ( ) NO
4. With which organisations or people?
5. Was there any training for those working on the construction of the underground dams? If YES, what was it?
6. What is the construction method?

   ( ) Manual ( ) Machines (backhoe, bulldozer, etc.)
7. Were any of the previous studies listed below carried out before construction began?

   ( ) Topographical survey

   ( ) Probing

   ( ) Particle size studies

   ( ) Analyses (soil, water)

   ( ) No previous study was carried out

   ( ) Other: ____
8. Was labour used? Specialised or local?
9. How many underground dams have been built? And which locations?
10. What material is used for the waterproofing septum?
11. What is the approximate cost per dam?
12. Underground dams were built in which of the following environments?

( ) Rivers, temporary streams

( ) Areas with sufficient slope for water to accumulate

APPENDIX B - Questionnaire applied to owners of underground dams

Questionnaire

Age:

Gender: Male ( ) Female ( )

Profession:

Underground dam location community:

1. Are you the owner of the land?

   ( ) YES ( ) NO

2. What is the purpose, use, of the underground dam on the property? Note: The question can have more than one answer

   ( ) Agriculture (subsistence, rainfed, family)

   ( ) Livestock

   ( ) To feed animals

   ( ) Fodder

   ( ) Human consumption

   ( ) Other:

3. Do you have an Amazon well?

   ( ) YES ( ) NO

4. If the answer to the previous question is YES, was it built before or after the construction of the underground dam?

   ( ) BEFORE ( ) AFTER

5. Has the underground dam shown results? What results?

NOTE: Only answer questions 6 to 10 if you built the underground dam privately.

6. Were any of the previous studies listed below carried out before construction began?

   ( ) Topographical survey

   ( ) Probing

   ( ) Particle size studies

   ( ) Analyses (soil, water)

( ) No previous study was carried out

( ) Other

7. Was labour used? Specialised or local?
8. What is the construction method?

( ) Manual ( ) Machines (backhoe, bulldozer, etc.)

9. What material is used for the waterproofing septum?
10. Underground dams were built in which of the following environments?

( ) Rivers, temporary streams

( ) Areas with sufficient slope for water to accumulate

APPENDIX C - List of Underground Dams built on the initiative of the Municipal Government

MUNICIPALITY	STATE	COMMUNITY	QUANTITIES
Good Success	PB	Sailor's bag	3
Good Success	PB	Córrego do Boi	1
Good Success	PB	Humaitá	1
Good Success	PB	Oiticica dos Belos	3
Good Success	PB	Oiticica dos Abrantes	1
Good Success	PB	Oiticica dos Timóteo	2
Good Success	PB	Serrinha	1
Good Success	PB	Low	2
Good Success	PB	Lamarão	1
Good Success	PB	Hidden	1
Good Success	PB	Riacho do Sítio	1
Good Success	PB	Olho D'aguinha	1
Good Success	PB	São Bento	1
Good Success	PB	Soledade Creek	1
Good Success	PB	Mutamba	2

Table 2: List of underground dams built in the municipality of Bom Sucesso/PB

Source: Author's own

APPENDIX D - List of privately built underground dams that are the subject of this study.

MUNICIPALITY	STATE	COMMUNITY	QUANTITIES
Good Success	PB	Sitio Bom Sucesso	1
Good Success	PB	Oiticica dos Belos	1
Good Success	PB	Wet Passage	1

Table 3: List of privately built underground dams that were the subject of this study.

Source: Author's own

Printed by Books on Demand GmbH, Norderstedt / Germany